Australian Geographic

Australia's Amazing Whales and Dolphins

Australia's Amazing Whales and Dolphins

First published in 2018. Reprinted in 2021.
Australian Geographic Holdings Pty Ltd
52-54 Turner Street
Redfern NSW 2016
editorial@ausgeo.com.au
australiangeographic.com.au

Funds from the sale of this book go to support the Australian Geographic Society, a not-for-profit organisation dedicated to sponsoring conservation and scientific projects, as well as adventures and expeditions.

Illustrations: Brett Jarrett
Designer: Mike Rossi
Creative director: Mike Ellott
Editor: Lauren Smith
Assistant editor: Rebecca Cotton
Print production: Katrina O'Brien

Australian Geographic
Managing Director: Jo Runciman
Editor-in-chief: Chrissie Goldrick

Contents

Introduction:

Marine mammals

MARINE MAMMALS have the same characteristics as land mammals, with one key exception: they have adapted to spend all or part of their lives in the ocean. They are classified into four groups: cetaceans (whales, dolphins, and porpoises), pinnipeds (seals, sea lions, and walruses), sirenians (manatees and dugongs), and marine fissipeds (polar bears and sea otters).

Australia is blessed with an abundance of riches when it comes to the marine mammals that grace our coastline or pass through our waters. No fewer than 45 cetaceans – whales, dolphins and porpoises – have been recorded around Australia, more than half of the world's species. Some of these are familiar and often-sighted animals, such as humpback whales, southern right and dwarf minke whales, while others are transients, or are found far offshore and are very rarely spotted by people, such as ginkgo-toothed beaked whales and dwarf sperm whales.

With their acrobatic antics, cheeky 'smiles' and evident intelligence, dolphins and porpoises are among the best loved of the world's sea creatures. Within the 22 species that have been recorded in Australian waters, some will be familiar to many people, such as the common bottlenose dolphin, which is often seen riding the bow waves of ships and boats. Others that are less well known include the unusual Australian snubfin dolphin and the rare Burrunan dolphin. There are sociable species, such as the short-beaked common dolphin, which can form pods containing in excess of 10,000 animals, and more solitary species, like the spectacled porpoise. That species is Australia's only true porpoise, differing from dolphins by their teeth, mouth and fins.

FUR SEAL: JUSTIN GILLIGAN/AG; DOLPHINS: GERARD SOURY/GETTY; ORCA: MICHAEL WEBERBERGER/GETTY

Baleen or toothed

The term 'whale' is an informal term that is used for the larger cetacean species, including both baleen whales (which filter food from the water using the bristled baleen plates attached to the roofs of their mouths) and toothed whales, which typically hunt fish and squid. Toothed whales (also called odontocetes) include dolphins, porpoises, sperm whales and beaked whales, as well as belugas and narwhals. Some common names for cetacean species are a bit misleading, such as the killer whale, which is actually a species of dolphin.

THAT'S AMAZING

Both marine and land mammals share these characteristics:

- *Warm-blooded*
- *Breathe air through lungs*
- *Have hair or fur*
- *Bear live young*
- *Nurse young with milk*

In focus:

Blue whale

Balaenoptera musculus

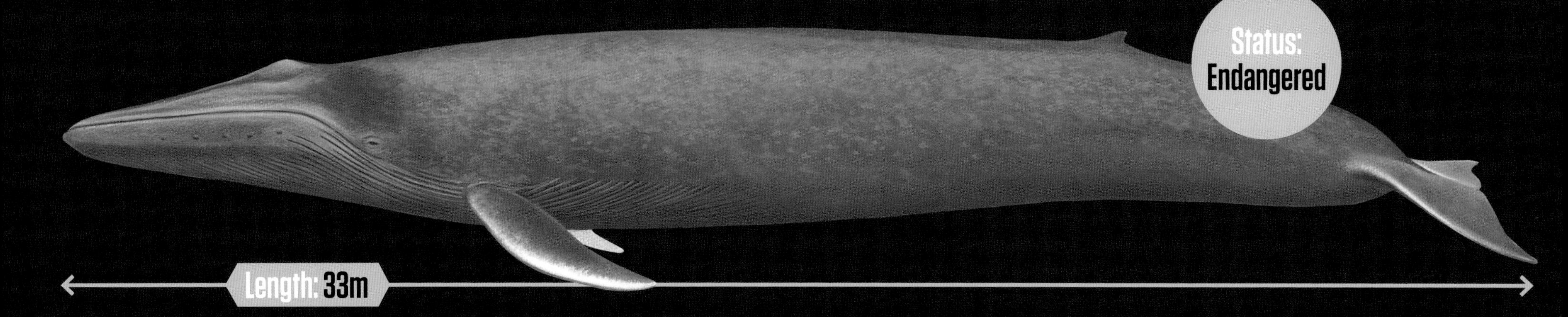

IT'S THOUGHT that no other animal that's ever lived on Earth has achieved the size of the blue whale, not even the biggest dinosaurs. The largest blue whale on record weighed almost 200 tonnes and was a length of more than 30m. They might be the biggest creatures on Earth, but blue whales prey only on tiny crustaceans called krill – up to 4 tonnes a day – which they strain from polar waters.

Blue whales live for 80–90 years. Decimated by 20th-century whalers, their numbers today are just 3–11 per cent of the 300,000 that once existed. An open-ocean species that often travels alone or in pairs, it can be seen off southern Australia. The subspecies found in Australia is the pygmy blue which reaches 25m in length.

LEFT: FRANCO BANFI/GETTY; DRONE SHOT: CHASE DEKKER/SHUTTERSTOCK (SS); ILLUSTRATION: MICHAEL PAYNE/AG

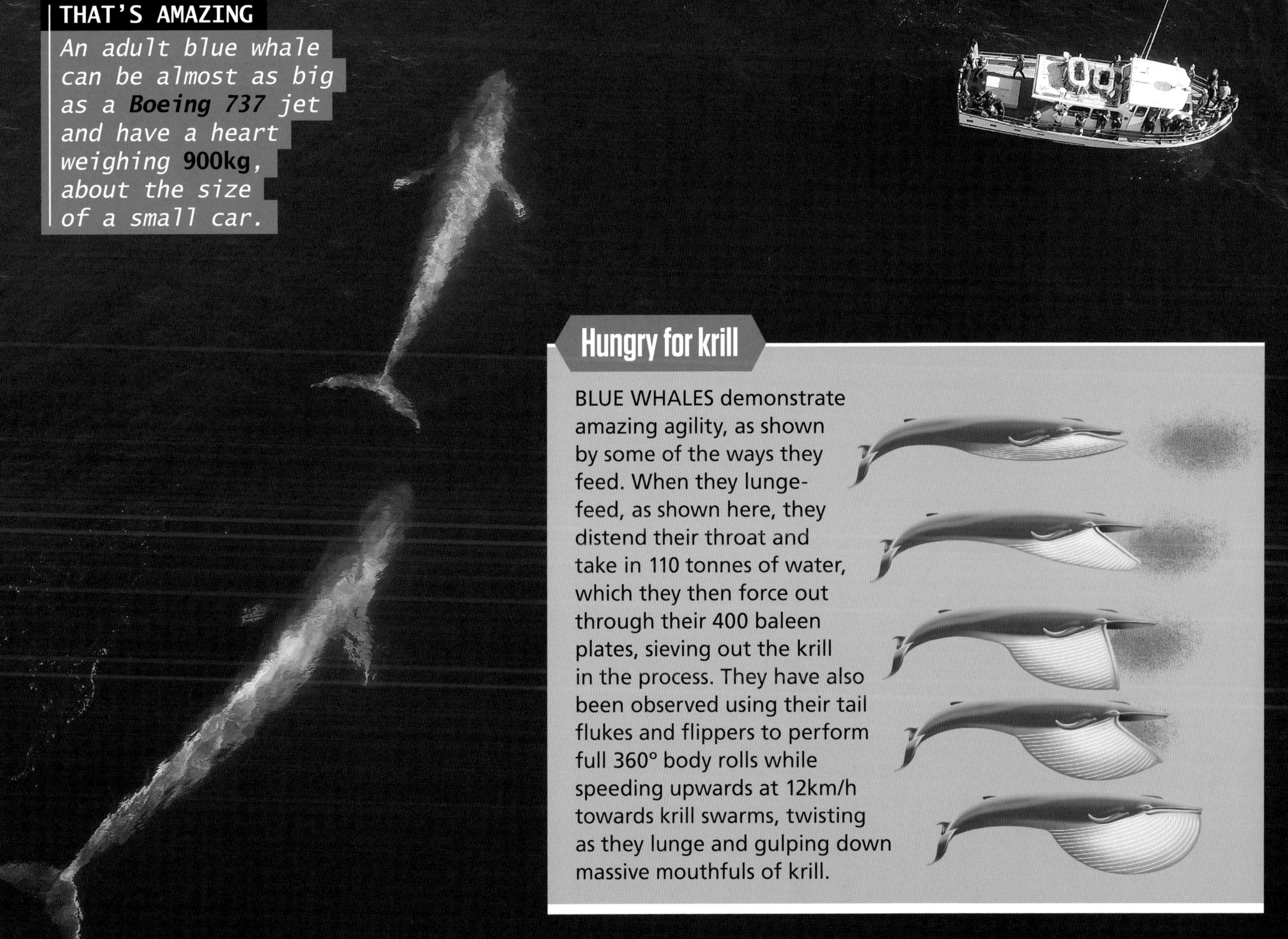

THAT'S AMAZING

An adult blue whale can be almost as big as a ***Boeing 737*** *jet and have a heart weighing* **900kg**, *about the size of a small car.*

Hungry for krill

BLUE WHALES demonstrate amazing agility, as shown by some of the ways they feed. When they lunge-feed, as shown here, they distend their throat and take in 110 tonnes of water, which they then force out through their 400 baleen plates, sieving out the krill in the process. They have also been observed using their tail flukes and flippers to perform full 360° body rolls while speeding upwards at 12km/h towards krill swarms, twisting as they lunge and gulping down massive mouthfuls of krill.

Baleen whales

Instead of teeth, baleen whales have long plates of baleen that hang like the teeth of a comb from their upper jaws, through which they filter huge amount of water to capture food. Baleen is made out of keratin, which is the strong, flexible material in human nails and hair.

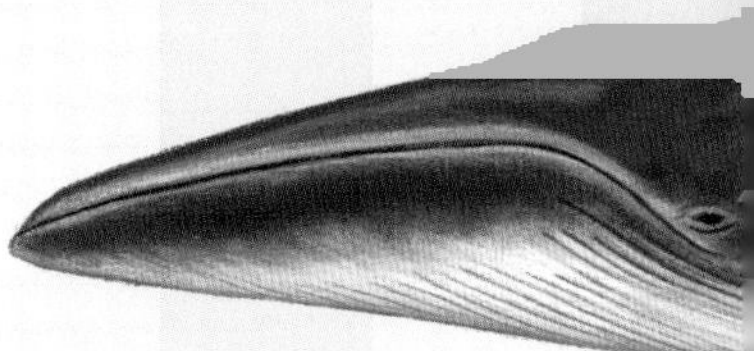

→

Sei whale

Balaenoptera borealis

Length: 12–18m **Status:** Endangered

Sei whales are among the fastest cetaceans, reaching speeds of more than 25km/h.

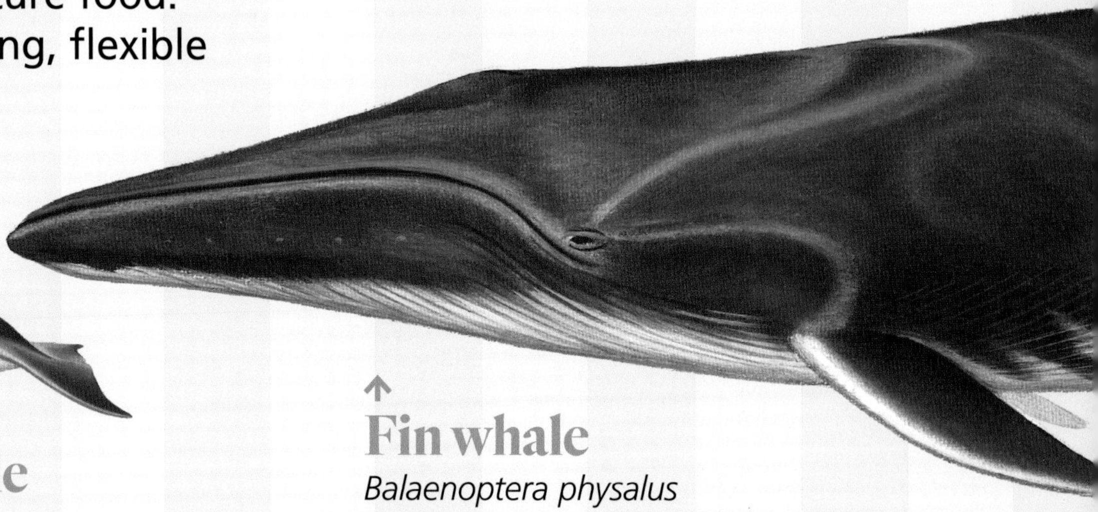

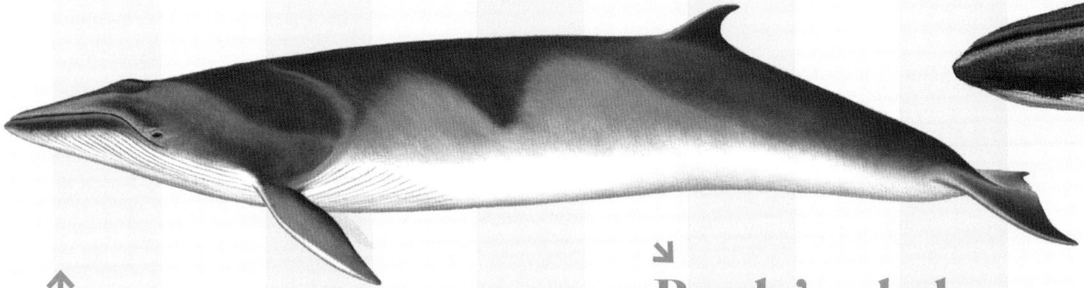

↑

Antarctic minke whale

Balaenoptera bonaerensis

Length: 8.5–10.7m **Status:** Data deficient

Antarctic minke whales feed on krill and small schooling fish. They are thought to be circumpolar, with most spending summer in Antarctic feeding grounds before migrating north to winter breeding grounds in Brazil, South Africa and Australia.

↘

Bryde's whale

Balaenoptera edeni

Length: 15m (m), 16.5m (f) **Status:** Data Deficient

Bryde's whales have three distinctive ridges on top of their heads. They are capable of diving to about 300m and swimming up to 25km/h, and use bubble nets to trap prey – mostly schooling fish.

↑

Fin whale

Balaenoptera physalus

Length: 27m **Status:** Endangered

Thought to reach speeds of up to 37km/h, the fin whale is not only the second-largest whale, but also one of the fastest. The only known predator of fin whales is the killer whale.

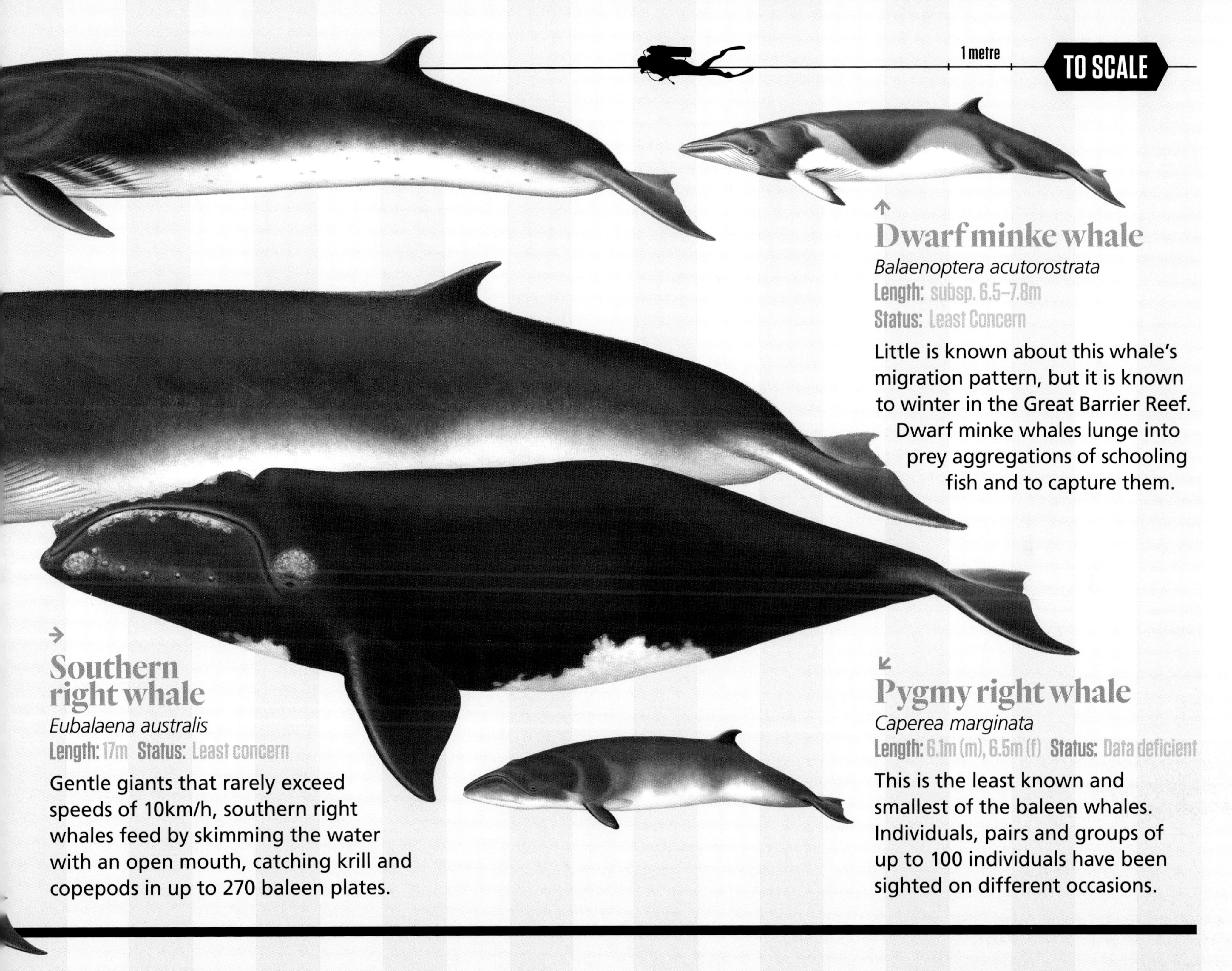

↑

Dwarf minke whale

Balaenoptera acutorostrata

Length: subsp. 6.5–7.8m

Status: Least Concern

Little is known about this whale's migration pattern, but it is known to winter in the Great Barrier Reef. Dwarf minke whales lunge into prey aggregations of schooling fish and to capture them.

→

Southern right whale

Eubalaena australis

Length: 17m **Status:** Least concern

Gentle giants that rarely exceed speeds of 10km/h, southern right whales feed by skimming the water with an open mouth, catching krill and copepods in up to 270 baleen plates.

↙

Pygmy right whale

Caperea marginata

Length: 6.1m (m), 6.5m (f) **Status:** Data deficient

This is the least known and smallest of the baleen whales. Individuals, pairs and groups of up to 100 individuals have been sighted on different occasions.

In focus:

Humpback whale

Megaptera novaeangliae

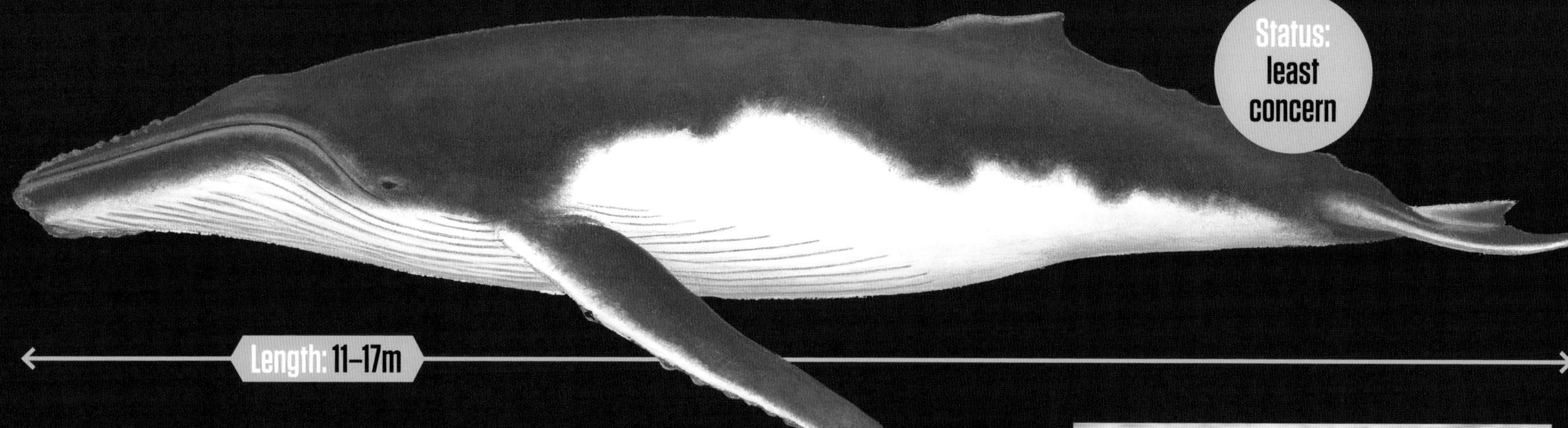

Status: least concern

Length: 11–17m

THERE ARE many humpback populations around the world, including those that migrate along Australia's coasts. They have a distinctive knobbly head, robust body shape, long flippers and are often covered in barnacles. Highly acrobatic, humpbacks can perform full breaches, levering their whole body out of the water. In the Southern Hemisphere they feed mainly on krill, sieving them from sea water using their baleen plates (right). They migrate yearly from feeding grounds in Antarctica waters to spend summer in warmer northern waters.

LEFT: JOHN TUNNEY/SS; RIGHT: ARTERRA/UIG/GETTY

THAT'S AMAZING

*During their migration these whales are known to travel as far as **16,000km**, making them one of the furthest migrating species in the world.*

Bubble net feeding

HUMPBACK WHALES have an extremely clever method of catching fish. Forming a team of up to a dozen whales, they 'trap' a school of fish by creating a circle of bubbles around them. The bubbles form a kind of net through which the fish cannot pass. The whales swim around and around, bringing their bubble net in tighter and tighter, closer and closer to the surface, until GULP! The whales lunge forward and take huge mouthfuls of fish in one go.

Toothed whales

As well as being distinguished by the presence of teeth, toothed whales also have a single blowhole (baleen whales have two) and are generally smaller than baleen whales. They hunt for their food, and some species are known to be quite aggressive.

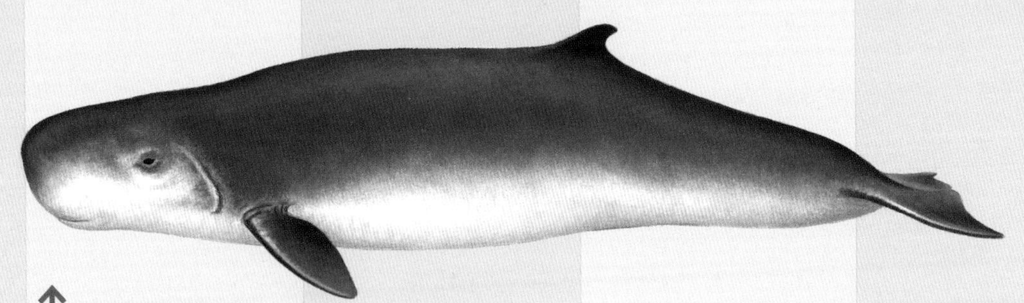

↑

Pygmy sperm whale

Kogia breviceps

Length: 2.7–3.8m **Status:** Data deficient

The pygmy sperm whale has an unusual shark-like head, with 12–16 pairs of sharp, fang-like teeth in its lower jaw, and 'false gills' on each side of its head, an adaptation thought to be related to the mimicry of sharks. They inhabit deep waters in warm oceanic zones worldwide.

↘

Dwarf sperm whale

Kogia sima

Length: 2.7m **Status:** Data deficient

Rarely sighted, it's thought these whales live in warm offshore waters, where they feed on a range of prey including deep-water cephalopods. Small pods of up to six are most common. They appear more dolphin-like than the pygmy sperm whale, and seem to prefer warmer waters.

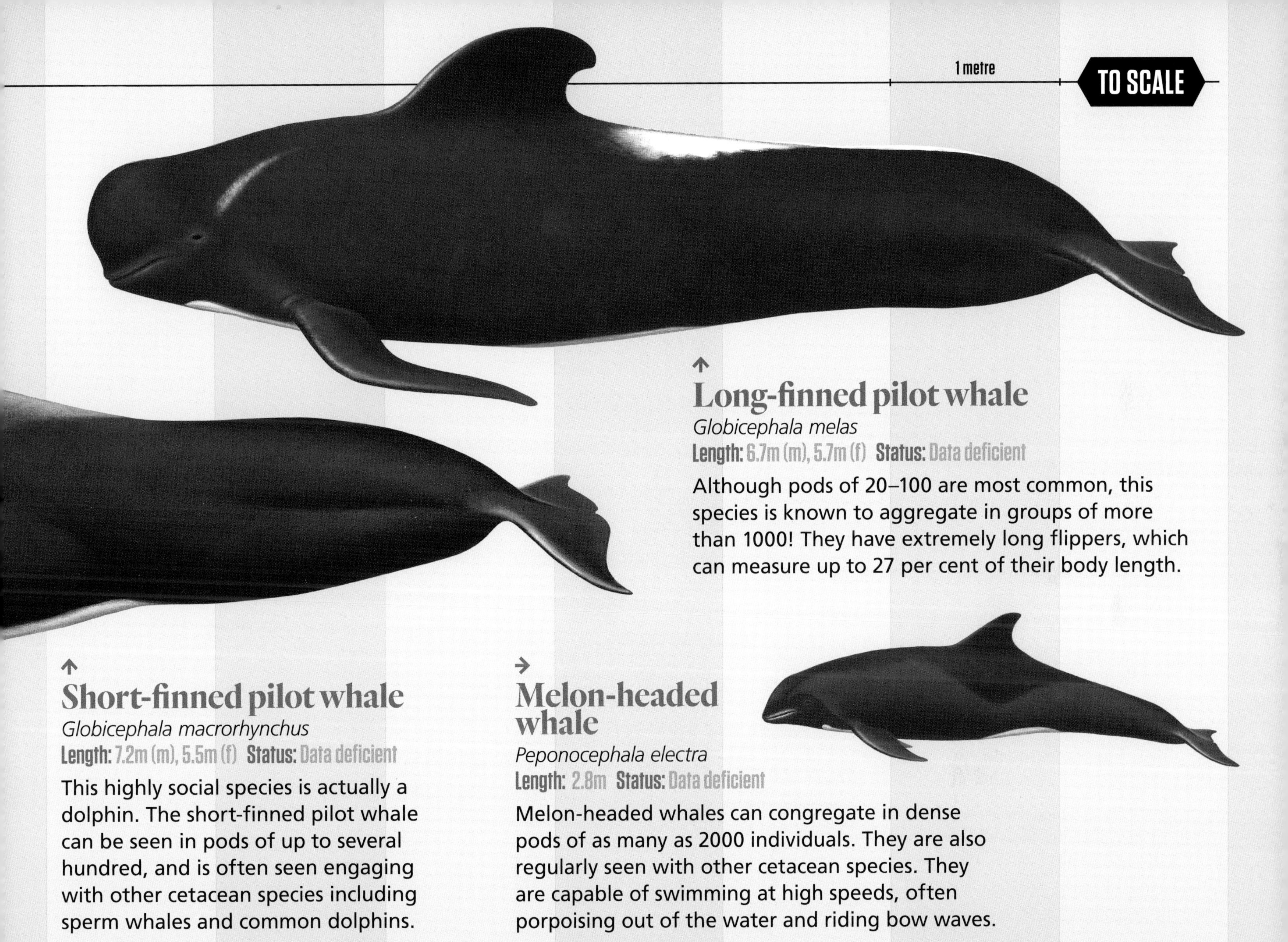

↑
Long-finned pilot whale

Globicephala melas

Length: 6.7m (m), 5.7m (f) **Status:** Data deficient

Although pods of 20–100 are most common, this species is known to aggregate in groups of more than 1000! They have extremely long flippers, which can measure up to 27 per cent of their body length.

↑
Short-finned pilot whale

Globicephala macrorhynchus

Length: 7.2m (m), 5.5m (f) **Status:** Data deficient

This highly social species is actually a dolphin. The short-finned pilot whale can be seen in pods of up to several hundred, and is often seen engaging with other cetacean species including sperm whales and common dolphins.

→
Melon-headed whale

Peponocephala electra

Length: 2.8m **Status:** Data deficient

Melon-headed whales can congregate in dense pods of as many as 2000 individuals. They are also regularly seen with other cetacean species. They are capable of swimming at high speeds, often porpoising out of the water and riding bow waves.

Squid snacking

It's estimated that giant and colossal squid comprise about 80 per cent of the sperm whale diet. Giant and colossal squid are the largest invertebrates alive today, and there have been reports of specimens up to 18m long washing up on beaches. These giant ceph-alopods aren't necessarily easy prey for sperm whales, and many sperm whales have scars on their backs that are thought to be caused by the sharp hooks in the suckers of the squid.

THAT'S AMAZING

Sperm whales nap vertically, with their heads pointed towards the ocean's surface, for about 10–15 minutes at a time.

THAT'S AMAZING

*The sperm whale has another extreme feature: it has an **8kg** brain, the largest of any animal that has ever lived.*

BACKGROUND: THE WILDEST ANIMAL/GETTY; SQUID: DORLING KINDERSLEY/GETTY

In focus:

Sperm whale

Physeter macrocephalus

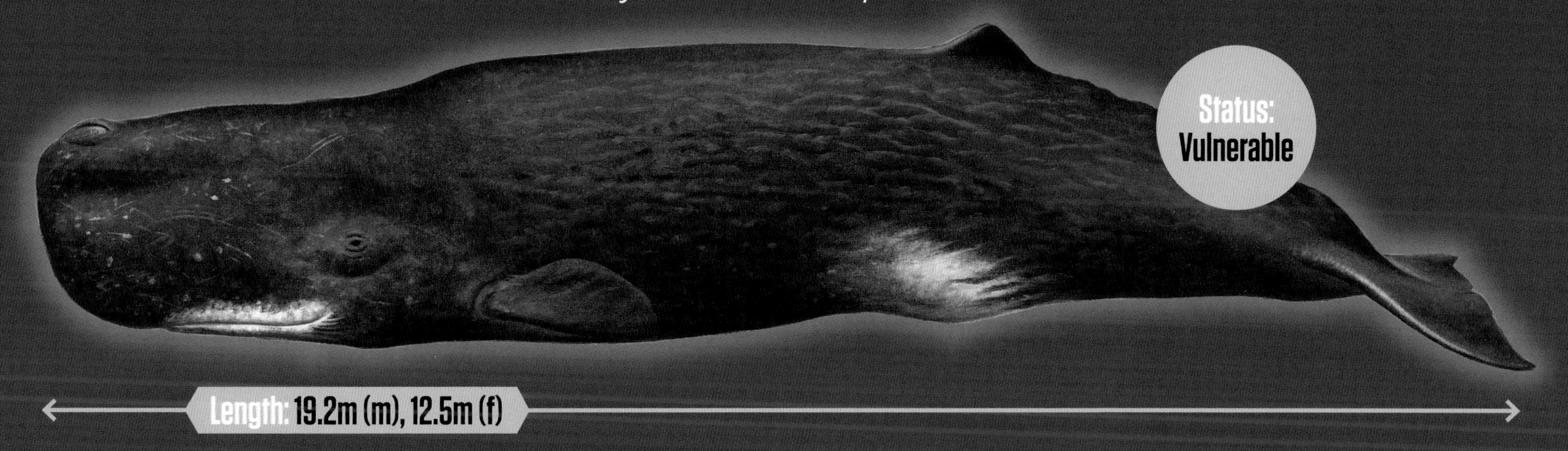

SPERM WHALES are the largest toothed cetaceans, and their compressed body shape and extremely large head make them readily distinguishable from other species. Pods of 20–30 are common, generally made up of adult females and their offspring. Males – which can be significantly larger than females – appear to associate with these groups for short periods to mate before moving on. They prey on deep-sea fish and cephalopods during 30–45 minute dives to depths of 400m or more.

They were once widely hunted for an oil they store in their heads, which led to a more than 60 per cent reduction in their population. This is activity is now largely banned around the world.

These huge deep ocean creatures appear to be able to communicate by producing clicking noises, the loudest of which have been measured at 230 decibels (dB). To give you an idea of how loud this is, a jet airplane taking off is about 165dB. Fortunately, sperm whale clicks are only short-lived – no more than 30 milliseconds – but they can be heard by another sperm whale more than 16km away.

Beaked whales

Beaked whales are distinguished by their elongated jaws, which form a 'beak'. This elusive group of whales has adapted to dive to great depths and as is one of the least known mammals groups to humans.

Blainville's beaked whale

Mesoplodon densirostris

Length: 4.7m **Status:** Data deficient

This whale is capable of diving to depths of up to 1400m, remaining submerged for nearly an hour.

Shepherd's beaked whale

Tasmacetus shepherdi

Length: 7m (m), 6.6m (f)

Status: Data deficient

The dark and light patches on this species make it distinguishable from other beaked whale species. They are thought to be an oceanic species, found in the Southern Hemisphere's deep, cold, temperate waters.

Strap-toothed beaked whale

Mesoplodon layardii

Length: 6.1m (m), 6.2m (f)

Status: Data deficient

Bizarre tusks protrude up to 30cm from the lower jaws of males, curling over the upper jaw and restricting the ability of some to open their mouths.

Anoux's beaked whale

Berardius arnuxii

Length: 9.3m **Status:** Data deficient

This species is generally found in groups of 6–10, but pods of up to 80 have been sighted. They are often covered in scratches from other whales and sharks.

Longman's beaked whale

Indopacetus pacificus

Length: 4–9m **Status:** Data deficient

Until 2003, this species was only known from skulls. They have been seen with spinner and bottle-nose dolphins and pilot whales.

1 metre

TO SCALE

↑

Southern bottlenose whale

Hyperoodon planifrons

Length: 7m (m), 7.5m (f)

Status: Data deficient

This species is found in oceanic waters beyond the continental shelf in the Southern Hemisphere, where it feeds on squid, fish and possibly crustaceans.

↑

Andrew's beaked whale

Mesoplodon bowdoini

Length: 4.4m **Status:** Data deficient

Having never been identified alive in the wild, this elusive species is only known from strandings, mainly in the South Pacific and Indian Oceans.

↑

True's beaked whale

Mesoplodon mirus

Length: 5.3–5.4m **Status:** Data deficient

The distinctions between the northern and southern forms of this species suggest they might be two species. Groups of up to three whales have been sighted.

→

Hector's beaked whale

Mesoplodon hectori

Length: 4.3m **Status:** Data deficient

Known only from strandings and a single sighting of a live animal, very little information has been able to be gathered about Hector's beaked whale.

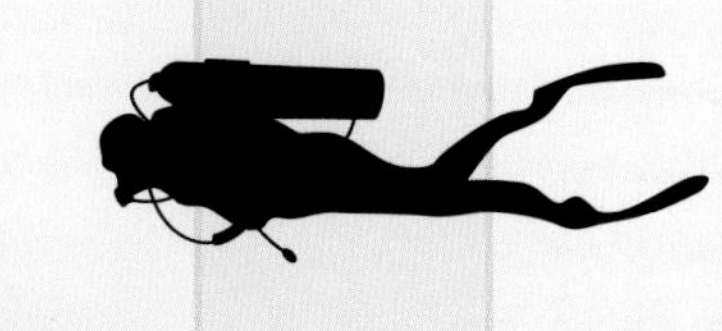

↑

Gray's beaked whale

Mesoplodon grayi

Length: 5.6m (m), 5.3m (f) **Status:** Data deficient

Sightings of single whales or pairs are most common, and they have been seen performing acrobatic breaches, as well as flipper- and fluke-slapping.

↑

Gingko-toothed beaked whale

Mesoplodon ginkgodens

Length: 5.3m **Status:** Data deficient

There have been no confirmed live sightings of this species. They have short, flattened tusks. Their name comes from the shape of the tusks of juveniles, which resemble the leaves of the ginkgo tree.

In focus:

Cuviers beaked whale

Ziphius cavirostris

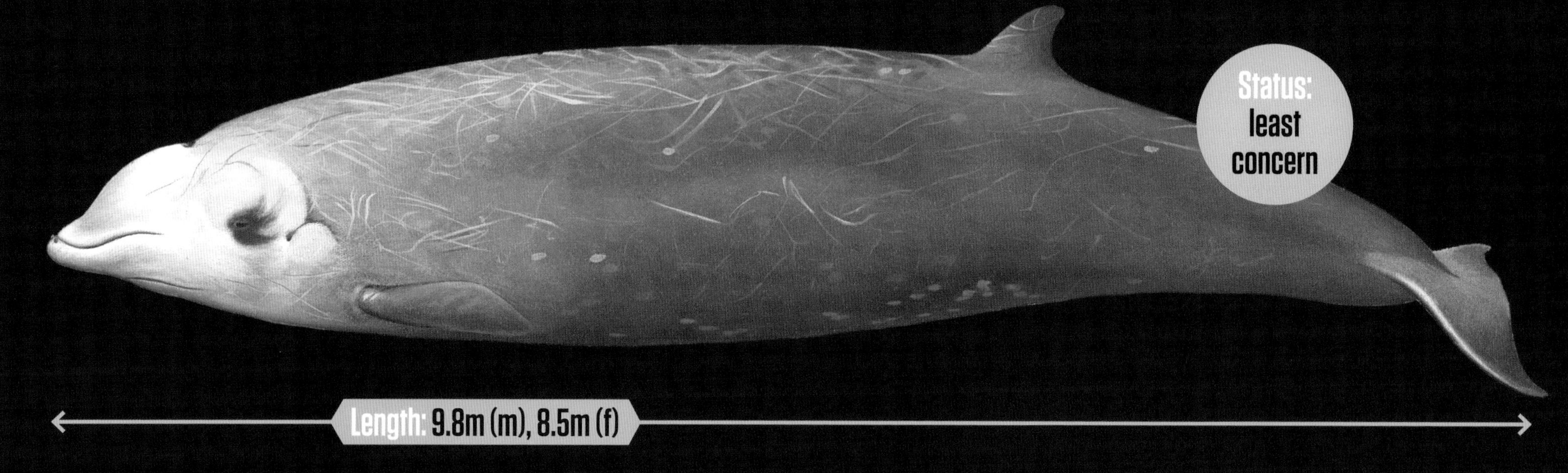

CUVIER'S BEAKED WHALES are the deepest divers of all marine mammals, and are capable of reaching quite incredible depths of nearly 3km and remaining submerged for more than two hours. The oval-shaped marks that often pepper adults' bodies are thought to be the result of bites from lampreys or cookie-cutter sharks. It's one of the most abundant of all of the beaked whales, being widely distributed throughout all oceans in both the Northern and Southern hemispheres. It's most commonly found in pods of 2–7 individuals.

INSET THIS PAGE: TODD PUSSER/GETTY; BACKGROUND OPPOSITE: ANDREA IZZOTTI/SS

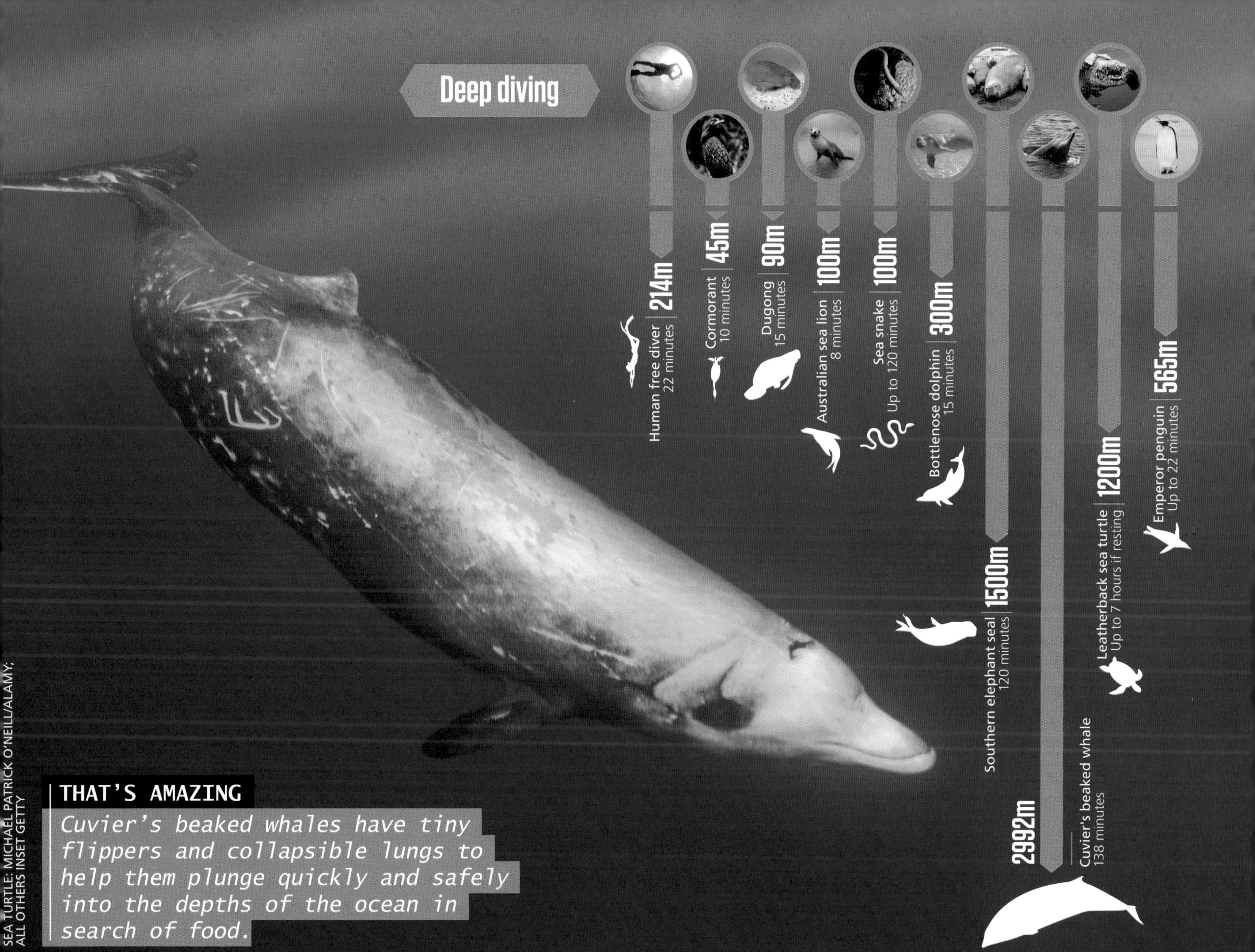

THAT'S AMAZING

Cuvier's beaked whales have tiny flippers and collapsible lungs to help them plunge quickly and safely into the depths of the ocean in search of food.

SEA TURTLE: MICHAEL PATRICK O'NEILL/ALAMY; ALL OTHERS INSET GETTY

Killer whales

The killer whales fall into a group of cetaceans known colloquially as 'blackfish' due to their predominantly dark-coloured bodies. The short- and long-finned pilot whales are also in this category. Killer whales are members of the Delphinidae family of cetaceans, which are often distinguinshed by a bulging forehead.

→

False killer whale

Pseudorca crassidens

Length: 6m (m), 5m (f) **Status:** Data deficient

With relatively uniform, dark-grey to black colouration, the false killer whale is the third-largest dolphin in existence. These deep oceanic animals are very social, generally found in pods of 10–60, ranging from extremely tight-knit groups to several smaller subgroups dispersed across more than 1km of ocean. Many mass strandings of this species have been documented, including one that was reported to involve more than 800 individuals.

Pygmy killer whale

Feresa attenuata

Length: 2.6m **Status:** Data deficient

This species is seldom sighted. It is dark grey to black in colour, often with white on the lips and tip of the beak. It's a relatively slow-swimming species found in groups of 12–50, often seen travelling in 'chorus lines', where individuals swim shoulder to shoulder, but pods as large as several hundred are known.

THAT'S AMAZING

The orca is the largest type of dolphin.

Helping hunters

Legend has it that orcas once helped human whalers to capture their marine victims off the coast of Eden, NSW. Between 1840 and 1930 the orcas helped whalers hunt baleen whales by herding the doomed giants towards the waiting whaleboats. Eyewitnesses talked of orcas prowling the entrance of Twofold Bay for migrating humpback, blue, southern right and minke whales. The waiting orcas would ambush whales that were vastly bigger than themselves, forcing them into shallower waters for the whalers to finish off, before claiming their piece of the prize.

BACKGROUND: TORY KALLMAN/SS; INSET: CHARLES EDEN WELLINGS/WIKIMEDIA COMMONS

In focus:

Orca

Orcinus orca

Status: Data deficient

Length: 9.8m (m), 8.5m (f)

ALSO KNOWN as killer whales, orcas are easily distinguished by their black-and-white colouration and large dorsal fin, which can reach more than 2m in males. They are known to aggregate in pods of up to 100, hunting a wide range of prey. These very intelligent hunters have mouths full of 8cm-long teeth. Orcas grow to weights of more than five tonnes and are able to reach speeds of almost 50km/h.

Killer whales in Argentina are known to capture newborn seals and sea lions right on the shore, timed perfectly to the ebb and flow of waves. In the Antarctic, orcas work together to knock seals off ice floes by swimming toward the ice and dipping down at the last minute to create a wave.

Pale and right whale dolphins

Twenty-two species of dolphin have been recorded in Australian waters, including pale dolphins and right whale dolphins. Some of these species are particularly rare, such as the elusive Burrunan dolphin known only from Australian waters.

Burrunan dolphin

Tursiops australis

Length: 2.6m **Status:** Not listed

Endemic to Australia, the entire population may be as small as a few hundred individuals, found in enclosed bays, estuarine systems and inshore coastal waters.

Indo-Pacific bottlenose dolphin

Tursiops aduncus

Length: 2.7m **Status:** Data deficient

This species is often found in pods of fewer than 20. In Australia it occurs right around the mainland, typically in estuarine and coastal waters. It feeds on a variety of fishes and cephalopods, and is regularly seen following fishing boats.

Rough-toothed dolphin

Steno bredanensis

Length: 2.8m **Status:** Least concern

Generally found in pods of 10–20, rough-toothed dolphins are well known for their habit of stealing bait and fish from fishing lines.

1 metre

TO SCALE

Australian humpback dolphin

Sousa sahulensis

Length: 2.7m

This species frequently enters estuaries and has been seen swimming up to 55km up rivers. Although it can occur in groups of up to 30, pods typically include only a few individuals.

THAT'S AMAZING

*Scientists studied the Australian humpback whale for **17 years** before deciding it was its own species in 2014.*

Risso's dolphin

Grampus griseus

Length: 3.8m **Status:** least concern

This robust, heavily-scarred, unusually shaped animal looks front-heavy, with a blunt head, bulging forehead and no distinctive beak. It's generally seen in pods of between 10 and several hundred individuals. It has been seen swimming in an 'echelon formation', lined up abreast at evenly spaced intervals.

Southern right whale dolphin

Lissodelphis peronii

Length: 3m **Status:** Data deficient

These energetic animals are commonly found in large pods made up of hundreds of individuals and can travel at speeds of up to 22km/h. It's thought capable of diving to more than 200m.

Common bottlenose dolphin

Tursiops truncatus

PERHAPS THE best-known and potentially most beloved dolphin species, the common bottlenose often lives in pods of fewer than 20 individuals in coastal waters, but offshore pods can number in the hundreds. They feed on a variety of prey using a range of strategies, including catching leaping fish mid-air and 'strand feeding', during which dolphins pursue and force fish onto a sandbar or beach, then launch from the water to feed on them. It's an active species known for energetic behaviour, which can include breaching and bow-riding.

INSET: TORY KALLMAN/SS; BACKGROUND: GABRIEL BARATHIEU/GETTY

THAT'S AMAZING

Bottlenose dolphins locate their prey by using ***echolocation****. They make clicking noises that travel through the water and bounce off objects, giving them information about the location and shape of the target. These dolphins can make up to* ***1000 clicking noises*** *per second.*

Best of friends

Just like humans, bottlenose dolphins are extremely social creatures that form strong bonds with other individuals that they are not necessarily related to. Research has shown adult male bottlenose dolphins form lifelong bonds with other select males for hunting and finding females. Members of an alliance typically spend up to 80-90 per cent of their time together and behave more like a group of friends than a hierarchical unit.

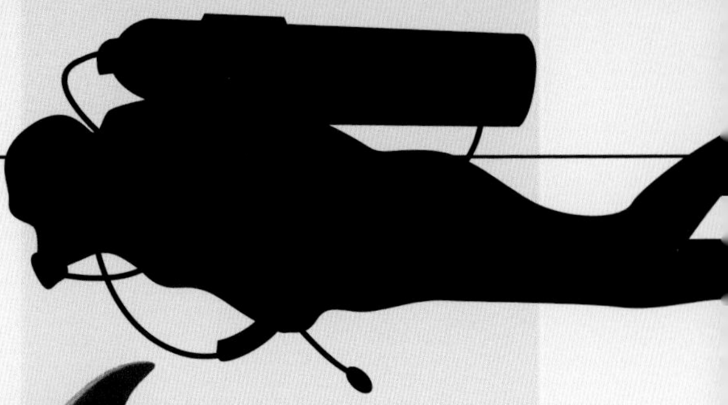

Narrow-beaked dolphins

Narrow-beaked dolphins include some of the most beautifully patterned of all the dolphin species.

Short-beaked common dolphin

Delphinus delphis

Length: 2.3m (m), 2.2m (f)
Status: Least concern

These energetic animals congregate in pods of 10–10,000 individuals, often swimming at high speed, performing acrobatic leaps and vocalising loudly.

Pantropical spotted dolphin

Stenella attenuata

Length: 2.6m (m), 2.4m (f) **Status:** Least concern

Adults of this slender species vary in the white mottling they have. Coastal animals typically have more spots than those living offshore, which can be completely spot-free.

Spinner dolphin

Stenella longirostris

Length: 2.4m (m), 2m (f)
Status: Data deficient

Named for its habit of leaping out of the water and spinning in the air, this species feeds on fish, crustaceans and cephalopods.

Striped dolphin

Stenella coeruleoalba

Length: 2.6m **Status:** Least concern

Striped dolphins appear to be more easily startled by boats than other dolphin species, and have been dubbed 'streakers' by fishermen. The species is very fast-swimming.

1 metre

TO SCALE

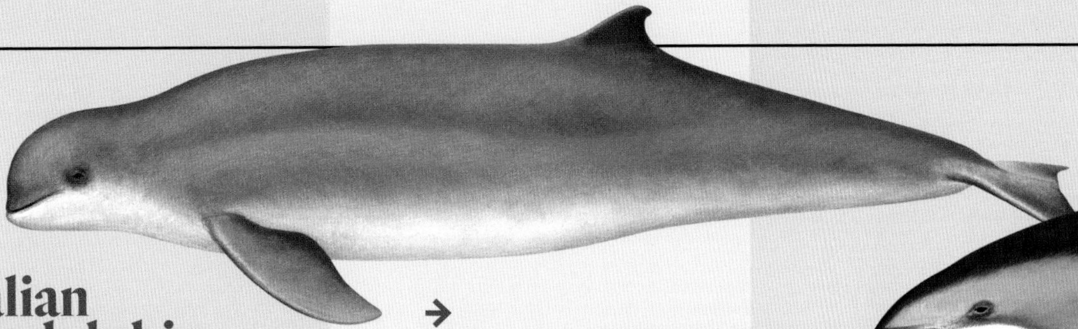

Australian snubfin dolphin

Orcaella heinsohni

Length: 2.7m vulnerable
Status: Least concern

Australian snubfins are shy creatures that may dive for long periods when startled and are usually seen in groups of five or six, although pods of up to 15 have been observed.

Hourglass dolphin

Lagenorhynchus cruciger

Length: 1.9m **Status:** Least concern

The hourglass dolphin is most often encountered in waters colder than 7°C, in pods of eight or fewer individuals.

Dusky dolphin

Lagenorhynchus obscurus

Length: 2.1m **Status:** Data deficient

A highly acrobatic species, in summer this social dolphin typically congregates in dense pods containing several hundred individuals. In winter, pods usually number fewer than 20 animals.

Fraser's dolphin

Lagenodelphis hosei

Length: 2.7m (m), 2.6m (f) **Status:** Data deficient

This oceanic species is often sighted in dense pods numbering in the hundreds or thousands, frequently with other cetacean species. It's thought to be a deep-diving species capable of reaching depths of 600m to feed on fishes, squid and crustaceans.

Short-beaked dolphins

While these groupings of dolphins are informal, these species are identifiable on the basis of their stubby beaks.

What's a porpoise?

At first glance it may be difficult to distinguish a porpoise from a dolphin, as they look very similar. However, the six species of porpoise that exist are generally smaller than dolphins, do not have a 'beak' and are rounder in body shape. They also posses spade-shaped teeth, compared to the conical teeth of dolphins. The spectacled porpoise (top) is Australia's only porpoise species. The harbour porpoise (middle) is found along coasts in the North Atlantic and North Pacific oceans, as well Arctic waters. The finless porpoise (bottom) inhabits the coastal waters of Asia.

THAT'S AMAZING

These marine mammals are excellent divers and head deep underwater to catch large fish and squid, helped by their large teeth. They are also known to feed on octopus, mantis shrimp, molluscs and other crustaceans.

DE AGOSTINI PICTURE LIBRARY/GETTY

In focus:

Spectacled porpoise

Phocoena dioptrica

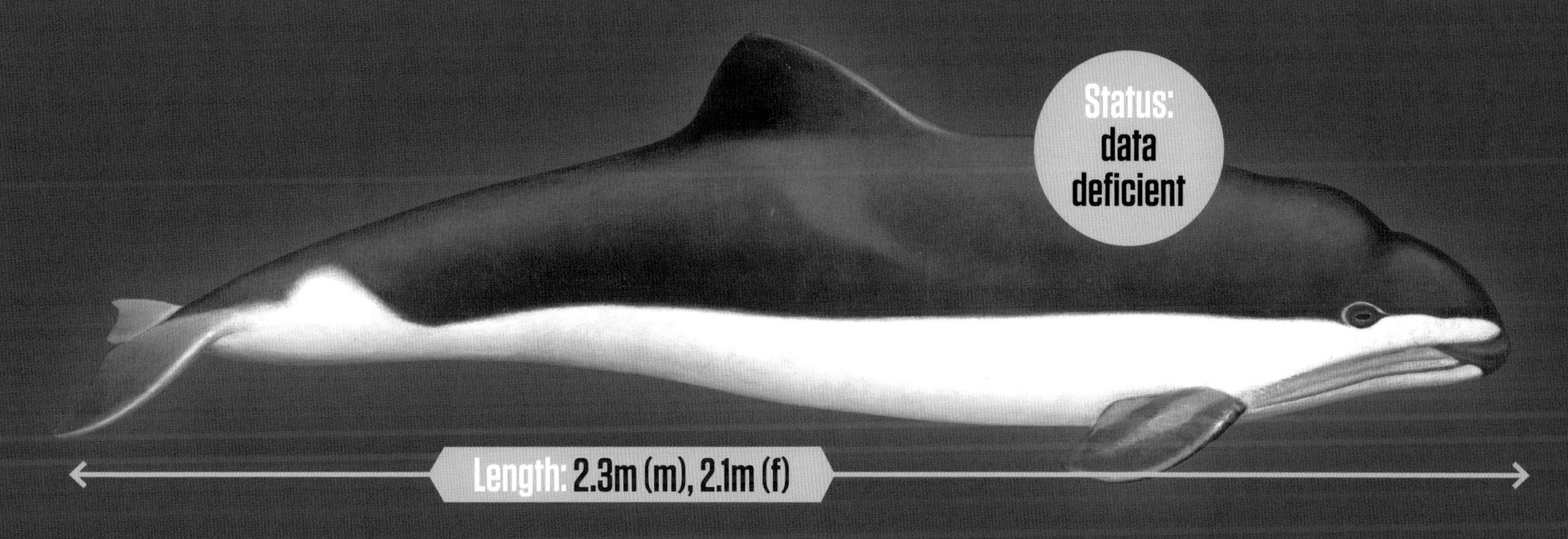

THIS DISTINCTIVE but rarely seen species is thought to live mainly in the freezing cold waters around Antarctica. Although believed to be an oceanic species, it's also been recorded in estuarine waters and is among the Southern Ocean's smallest cetaceans. Adults have heavily contrasting black-and-white colouration, with males possessing a large, oval-shaped dorsal fin that can look disproportionate to their bodies. This porpoise gets its name due to a black mark around the eye that is also circled in white, creating a 'spectacled' look. Individuals are typically seen alone, in pairs or in groups of three. They are known to be fast swimmers, and though they aren't particularly acrobatic aniamls, they do demonstrate some classic porpoising behaviour, including slow rolls on the surface and arching their backs when diving below the surface of the water. They are rarely seen, and not many photographs of live spectacled porpoises exist.

Saving our sea creatures

A NUMBER OF Australian marine mammals face the threat of extinction, predominantly as a result of human activity. Between the 18th and 20th centuries, whales were hunted for their meat, skin, baleen and organs. Whaling is now largely banned but some countries continue to hunt. Whales and dolphins are also at risk of becoming entangled in fishing gear and shark nets, colliding with boats and also face problems of pollution and debris. Climate change and the rising temperature of ocean waters mean that food sources for marine mammals are also dying out. There are five whale species currently considered threatened in Australia – the blue, southern right, sei, fin and humpback whales. Numbers of the snubfin dolphin are at critical levels, and populations for other dolphin species, such as the Burrunan dolphin, are extremly low. Here are some tips for helping our sea creatures.

ALL INSET: SHUTTERSTOCK